Ice-Age Monsters

Cave Bear

Written by Rupert Oliver
Illustrated by Bernard Long

Library of Congress Cataloging in Publication Data

Oliver, Rupert.
Cave Bear.

Summary: Follows a prehistoric cave bear through her day as she emerges from hibernation and introduces her new cub to the wonders and dangers of their savage world.
1. Cave bears—Juvenile literature. [1. Cave bears. 2. Bears. 3. Mammals, Fossil. 4. Prehistoric animals] I. Tong, Bernard, ill. II. Title.
QE882.C15043 1986 569'.74 86-577
ISBN 0-86592-844-4

Rourke Enterprises, Inc.
Vero Beach, FL 32964

Megaceros
Saber Tooth Tiger
Cave Bear

Ice-Age Monsters

Cave Bear

Ursus moved about restlessly. The weather was getting warmer and Ursus wanted to get outside and move around. She had been safe and warm in her small cave since before the first snows fell during the previous Autumn. While the snow lay thick outside and the wind howled coldly across the bleak landscape, Ursus had been sleeping in her den.

She had not been asleep the whole time, however. About half way through the Winter, Ursus had given birth to a baby. At first the baby had been small and helpless and had very little fur. During the remaining months of Winter, the baby had suckled milk from its mother and it had grown. Now the cub was quite large and covered with rich, thick fur.

Ursus smelled the air. It had the scent of fresh growth and living things. She stood up and ambled to the mouth of the cave. It had been many months since Ursus had left the cave, but today was different. Ursus looked out through the cave entrance at the surrounding hills. The snow had nearly all melted. Large stretches of greenery spread up the hillsides where fresh grass was growing. Only a few patches of snow remained.

Not very far away, a group of reindeer were grazing contentedly on some grass. Ursus decided that it was time to leave her winter lair and find some food.

After calling gruffly to her cub to follow her, Ursus walked out of the den. The cub had never been outside before. He leaped and jumped around in the Spring sunshine, but he was careful to stay close to his mother. As the two bears moved down the hillside, reindeer trotted off a short distance and then began eating again.

When the bears reached the valley floor, Ursus looked around. Soon, she grunted with satisfaction and walked across the valley until she came to a patch of green shoots pushing up through the grass. Ursus began to dig in the ground. A few inches below the shoots were bulbs. As soon as she had dug up a few bulbs, Ursus began to eat them. These bulbs were tasty and full of goodness. When the cub saw his mother eating the bulbs he thought he would try one. The cub took a small bite. The bulb tasted nice, but not as nice as his mother's milk.

Ursus bent her head down to drink the ice cold water. She lapped up the water greedily. The cub did the same and found that he liked the cold water.

Suddenly, a roar boomed across the river. Ursus looked up in alarm. Perhaps there was danger. On the far side of the river a large Homotherium was attacking Megaceros. The cat had been hiding in the forest undergrowth and had leaped out at the Megaceros. The Megaceros did not have time to defend itself and soon the sharp teeth of the Homotherium were tearing into the hapless victim. Ursus decided to take her cub away from the fight and the two bears moved back into the forest.

Ursus and her cub spent most of the day in the forest. Ursus ate her fill of leaves and bulbs. When she found a bush covered with the fresh shoots of Spring, Ursus hungrily stripped the shoots. They were her favorite food. The cub always stayed close to his mother since he knew this was the safest thing to do.

It was late in the afternoon when Ursus smelled something very strange. It was the smell of roasting meat. That smell often meant that men were nearby and men were dangerous. Ursus pushed her cub up into a tree and moved off to investigate. She did not see the Cave Lion hiding in the bushes.

Carefully, Ursus pushed her way through the undergrowth. If there were men nearby she would have to take her cub to another part of the forest for safety. Then she heard a cry of distress from behind her. It was her cub and he was in danger.

Ursus turned around and charged through the undergrowth toward her cub. When she came within sight of her cub, Ursus saw the Cave Lion trying to climb the tree where the cub was hiding. Without stopping, Ursus charged headlong at the Cave Lion. The Cave Lion turned just in time to see the advancing bulk of Ursus. Roaring its fear, the Cave Lion turned and ran. It did not want to fight a fully grown bear.

Ursus chased the Cave Lion for some distance through the forest. She wanted to be sure that the lion would not come back. When Ursus was certain her cub would not be in danger from the Cave Lion she returned through the forest.

Calling gently, the cub climbed down from the tree and came over to its mother. Ursus nuzzled her cub. She was glad he was safe. Night was beginning to fall in the forest and Ursus and her cub were tired. Ursus pushed her way into a very thick patch of undergrowth and called to her cub to follow. The two bears lay down in the thicket and slept soundly as night fell around them.

Cave Bear and the Pleistocene Ice Ages

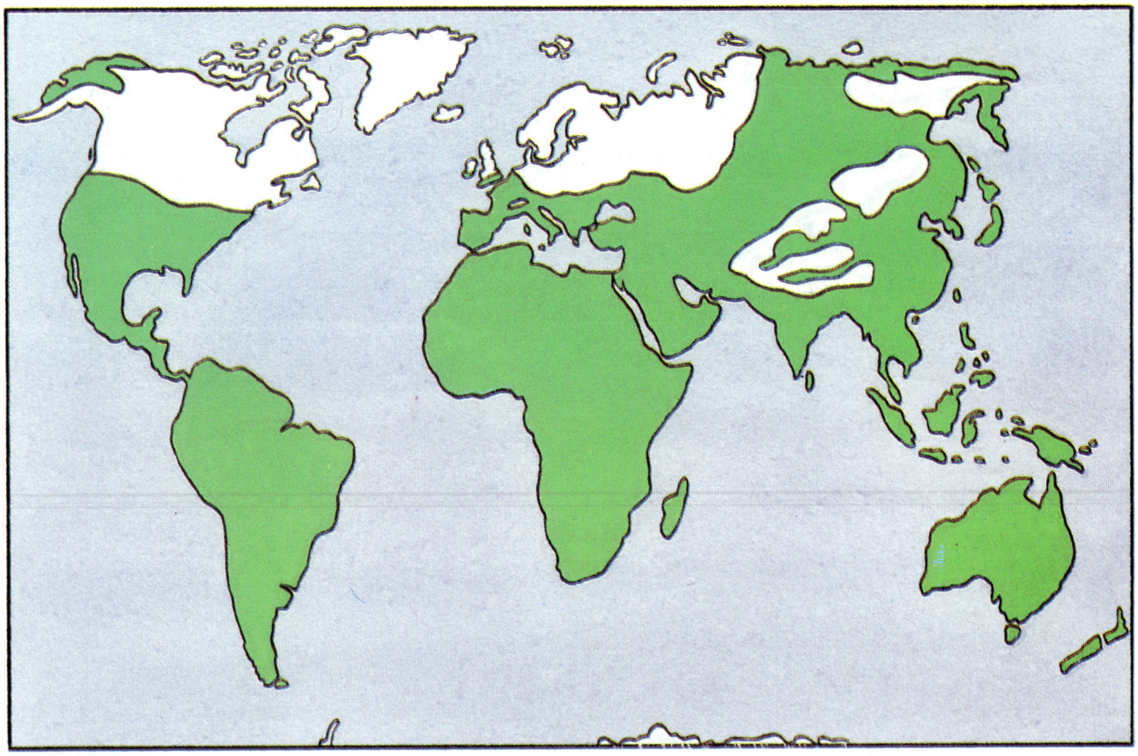

The Earth 50,000 years ago, showing the extent of the polar ice sheets.

The Ice Ages

Today, the earth is divided into a number of clearly defined climatic regions. Around the equator the climate is generally hot and moist. This gives rise to a wide stretch of jungle where tropical plants and animals thrive. On either side of the jungle region, extends land where deserts are very common and life is scarce. Nearer to the poles are the temperate zones. In these areas, deciduous forests, such as those of New England and Western Europe, are common, together with open grasslands like the Great Plains of North America. Nearer to the poles are broad belts of coniferous forest and tundra. Tundra is land which has such a cold climate that only small bushes, grass and lichens can survive. At the North and South poles the snow never melts in the Summer and great sheets of ice cover land and sea alike. Several times in the past million years, the climate of the entire world has become much colder and then much warmer. During the cold periods, the polar ice sheets reached far to the south of their present position. In America, they covered nearly all of Canada and reached the Great Lakes. In Europe they came as far south as London and northern Germany. During such times all the climatic zones of the world shifted south with tundra extending across much of North America. During the warm periods the polar caps almost disappeared and the zones moved north. Our story takes place during one of the cold periods about 50,000 years ago.

The Cave Bear

The species of bear in our story is the Cave Bear, which scientists call Ursus Spelaeus. The Cave Bear is now extinct; the last one died about 20,000 years ago. The Cave Bear acquired its name because so many of its bones have been found in caves throughout Europe and southwestern Asia. This does not mean that it lived in caves all its life. Like modern bears in cool climates, the Cave Bear hibernated during the Winter when there was not much food about. All Summer the bear would eat as much food as it could. This would build up thick layers of fat underneath the skin on which the bear would live during the Winter when it did not eat or drink. When the bear came out of its den in the Spring it would probably be very hungry and thirsty. It was during the Winter that female bears would give birth to their young. During the Winter the mother bear would suckle her cub so that it would be ready to follow her outside in the Spring. The cold weather of Winter was a dangerous time for Cave Bears. If a bear was not in a warm enough den the cold might kill it, while if it did not have enough fat it might starve to death. Because so many bears died during the Winter their bones were deposited in caves. Over the years many bears might die in the same den. In one cave in Austria 30,000 Cave Bear skeletons were found. Despite their fearsome appearance and great strength, Cave Bears were probably not great meat eaters. It seems that their favorite food was plants. They would eat leaves, shoots, roots, bulbs and even grass on occasions. Perhaps, like modern bears, the Cave Bears could catch fish in shallow rivers.

Ice Age Animals

During the long cold periods of the Ice Ages many animals evolved specially to survive in cold conditions. Since the cold climate could last for several thousand years and spread across most of the world, these animals became very common. The Cave Bear managed to survive by staying in a den during the worst of the weather, but other animals stayed outside. The most famous animal of the Ice Ages was the Woolly Mammoth. The way this animal coped with cold weather is clear. It evolved a thick coat of fur to keep itself warm. The outer layer of fur was long and coarse to keep the rain and snow out, while the inner fur was rich and soft to retain the warmth. Even with this protection, the Woolly Mammoth found Ice Age Winters severe. It is probable that during the Winter the herds of Woolly Mammoth moved South into the forest to escape a Winter on the tundra. The herd in our story were migrating North as Spring warmed the tundra. The Megaceros was another animal which spent the Summers on the open tundra, but which probably moved South for the Winter. Despite its name it was not really an elk, but a distinct species of giant deer. The most notable feature about the Megaceros was the size of its antlers. They were enormous, measuring more than 12 feet across. At the time of our story the Homotherium was a very rare animal. It was a large cat with long stabbing canines and was related to the Saber Tooth Tiger. Several thousand years before our story, it had been a common hunter throughout Europe and Asia, but by the time of the Cave Bear it was on the verge of extinction. The Cave Lion, on the other hand, was becoming increasingly common and was one of the deadliest hunters of the time.

The Cave Bear and Man

At the time of our story early man was living throughout Europe and Asia, where the Cave Bear lived. At this point man was living in the Old Stone Age. He made tools of stone and wood and lived by hunting wild animals and gathering wild plants. Early man seems to have had a strange relationship with the Cave Bear. In several caves which were inhabited by man, scientists have found Cave Bear skulls buried in special pits and arranged on shelves on the cave walls. Nobody knows why prehistoric man did this. Perhaps he was displaying hunting trophies or perhaps the skulls formed part of some strange ceremony. Either way it proves that man hunted Cave Bears and the bear in our story was wise to be frightened of them.

Two modern relatives of the Cave Bear

Black Bear

Grizzly Bear